GAO

Report to the Committee on Science, Space, and Technology, House of Representatives

I0471918

June 2012

POLAR-ORBITING ENVIRONMENTAL SATELLITES

Changing Requirements, Technical Issues, and Looming Data Gaps Require Focused Attention

G A O

Accountability ★ Integrity ★ Reliability

Highlights

Highlights of GAO-12-604, a report to the Committee on Science, Space, and Technology, House of Representatives

POLAR-ORBITING ENVIRONMENTAL SATELLITES

Changing Requirements, Technical Issues, and Looming Data Gaps Require Focused Attention

Why GAO Did This Study

Environmental satellites provide critical data used in forecasting weather and measuring variations in climate over time. NPOESS—a program managed by NOAA, DOD, and the National Aeronautics and Space Administration—was planned to replace two existing polar-orbiting environmental satellite systems. However, 8 years after a development contract for the NPOESS program was awarded in 2002, the cost estimate had more than doubled—to about $15 billion, launch dates had been delayed by over 5 years, significant functionality had been removed from the program, and the program's tri-agency management structure had proven to be ineffective. In February 2010, a presidential task force decided to disband NPOESS and, instead, to have NOAA and DOD undertake separate acquisitions.

GAO was asked to evaluate (1) efforts to transfer responsibilities from the NPOESS program to the separate NOAA and DOD programs, (2) NOAA's progress in developing its satellite system, and (3) NOAA's efforts to mitigate key project risks. To do so, GAO analyzed program management, contract, cost, and risk data, attended executive program reviews, and interviewed agency and contractor officials.

What GAO Recommends

GAO is making a recommendation to NOAA to establish mitigation plans to address the risk of satellite data gaps. NOAA agreed with GAO's recommendation and noted that the agency is developing a report to address the risk of data gaps.

View GAO-12-604. For more information, contact David Powner at (202) 512-9286 or pownerd@gao.gov.

What GAO Found

Following the decision to disband the National Polar-orbiting Operational Environmental Satellite System (NPOESS) program in 2010, both the National Oceanic and Atmospheric Administration (NOAA) and the Department of Defense (DOD) made initial progress in transferring key management responsibilities to their separate program offices. Specifically, NOAA established a Joint Polar Satellite System (JPSS) program office, documented its requirements, and transferred existing contracts for earth-observing sensors to the new program. DOD established its Defense Weather Satellite System program office and modified contracts accordingly. However, recent events have resulted in major program changes at both agencies. NOAA plans to revise its program requirements to remove key elements, including sensors and ground-based data processing systems, to keep the program within budget. Further, in early 2012, DOD decided to terminate its program and reassess its requirements.

Over the past year, NOAA has made progress in developing its satellite system, but critical decisions and milestones lie ahead. In October 2011, the JPSS program office successfully launched a satellite originally called the NPOESS Preparatory Project (NPP). Data from the satellite are currently being calibrated and validated, and NOAA meteorologists started using selected satellite data products in their weather forecasts in May 2012. Further, the three major components of the JPSS program (the flight, ground, and free-flyer projects) are at different stages of development. Within the flight project, development of the sensors for the first JPSS satellite is well under way; however, selected sensors are experiencing technical issues. The ground project is currently in operation supporting NPP and NOAA is planning to upgrade parts of the ground system infrastructure to increase its security and reliability. The free-flyer project, intended to integrate and launch key instruments that could not be accommodated on the JPSS satellites, is still in a planning stage because NOAA has not yet decided which satellites will host the instruments or when these satellites will launch.

The JPSS program office has implemented elements of an effective risk management process; however, the program still faces significant risks. It does not yet have a cost and schedule baseline in place, the program office is not yet fully staffed, and there will likely be a gap in satellite data lasting 17 to 53 months from the time NPP is projected to cease operations and the first JPSS satellite begins to operate. There are also potential satellite data gaps in the DOD and European polar satellite programs, which provide supplementary information to NOAA forecasts. The JPSS program office is managing the first two risks, but NOAA has not established plans to mitigate potential satellite gaps. Until these risks are mitigated and resolved, civilian and military satellite data users may not have the information they need for timely weather forecasting, thereby risking lives, property, and commerce.

Contents

Abbreviations

ATMS	Advanced Technology Microwave Sounder
CERES	Cloud and Earth's Radiant Energy System
CrIS	Cross-Track Infrared Sounder
DMSP	Defense Meteorological Satellite Program
DOD	Department of Defense
DWSS	Defense Weather Satellite System
IDPS	Interface Data Processing Segment
JPSS	Joint Polar Satellite System
MetOp	Meteorological Operational (satellite)
NASA	National Aeronautics and Space Administration
NOAA	National Oceanic and Atmospheric Administration
NPOESS	National Polar-orbiting Operational Environmental Satellite System
NPP	NPOESS Preparatory Project (original)/ National Polar-orbiting Partnership (current)
OMPS	Ozone Mapping and Profiler Suite
POES	Polar-orbiting Operational Environmental Satellites
VIIRS	Visible/Infrared Imager/Radiometer Suite

United States Government Accountability Office
Washington, DC 20548

June 15, 2012

The Honorable Ralph Hall
Chairman
The Honorable Eddie Bernice Johnson
Ranking Member
Committee on Science, Space, and Technology
House of Representatives

The National Polar-orbiting Operational Environmental Satellite System (NPOESS) program was planned to be a state-of-the-art, environment-monitoring satellite system that would replace two existing polar-orbiting environmental satellite systems. Managed jointly by the Department of Commerce's National Oceanic and Atmospheric Administration (NOAA), the Department of Defense (DOD)/U.S. Air Force, and the National Aeronautics and Space Administration (NASA), the program was considered critical to the nation's ability to maintain the continuity of data required for weather forecasting and global climate monitoring through the year 2026.

However, in the 8 years after the development contract was awarded in 2002, the NPOESS cost estimate had more than doubled—to about $15 billion, launch dates had been delayed by over 5 years, significant functionality had been removed from the program, and the program's tri-agency management structure had proven to be ineffective. Importantly, delays in launching the satellites put the program's mission at risk. To address these challenges, a task force led by the White House's Office of Science and Technology Policy reviewed the management and governance of the NPOESS program. In February 2010, the Director of the Office of Science and Technology Policy announced a decision to disband the NPOESS acquisition and, instead, to have NOAA and DOD undertake separate acquisitions, with NOAA responsible for satellites in the afternoon orbit and DOD responsible for satellites in the early morning orbit. After that decision, both agencies began developing plans for their separate programs, called the Joint Polar Satellite System (JPSS) and the Defense Weather Satellite System (DWSS), respectively.

This report responds to your request that we (1) evaluate efforts to transfer management and contract responsibilities from the NPOESS program to the separate NOAA and DOD programs, (2) assess NOAA's progress in developing the NPOESS Preparatory Project (NPP) and JPSS, and (3) evaluate NOAA's efforts to mitigate key project risks.

To evaluate efforts to transfer management and contract responsibilities from NPOESS programs to separate NOAA and DOD programs, we compared NOAA's and DOD's plans for establishing program management offices and transferring contracts to each agency's actual accomplishments. We also observed NOAA's monthly program management briefings and interviewed NOAA, NASA, and DOD officials to obtain insights into risks, issues, and transition schedules. To assess progress in developing the NPP and JPSS satellite systems, we compared NOAA's plans for key milestones to its accomplishments and interviewed agency and contractor officials. We also interviewed key NOAA and DOD satellite data users to determine their experiences in working with NPP data as well as their plans for working with JPSS data. To evaluate NOAA's efforts to mitigate key project risks, we compared the agency's risk management process to best practices in risk management as identified by leading systems engineering organizations, and interviewed agency officials.

We conducted this performance audit from May 2011 to June 2012 in accordance with generally accepted government auditing standards. Those standards require that we plan and perform the audit to obtain sufficient, appropriate evidence to provide a reasonable basis for our findings and conclusions based on our audit objectives. Additional details on our objectives, scope, and methodology are provided in appendix I.

Background

Since the 1960s, the United States has operated two separate operational polar-orbiting meteorological satellite systems: the Polar-orbiting Operational Environmental Satellite (POES) series, which is managed by NOAA, and the Defense Meteorological Satellite Program (DMSP), which is managed by the Air Force.[1] These satellites obtain environmental data that are processed to provide graphical weather images and specialized weather products. These satellite data are also the predominant input to numerical weather prediction models, which are a primary tool for forecasting weather days in advance—including forecasting the path and intensity of hurricanes. The weather products and models are used to predict the potential impact of severe weather so that communities and emergency managers can help prevent and

---

[1]NOAA provides command and control for both the POES and DMSP satellites after they are in orbit.

mitigate its effects. Polar satellites also provide data used to monitor environmental phenomena, such as ozone depletion and drought conditions, as well as data sets that are used by researchers for a variety of studies such as climate monitoring.

Unlike geostationary satellites, which maintain a fixed position relative to the earth, polar-orbiting satellites constantly circle the earth in an almost north-south orbit, providing global coverage of conditions that affect the weather and climate. Each satellite makes about 14 orbits a day. As the earth rotates beneath it, each satellite views the entire earth's surface twice a day. Currently, there is one operational POES satellite and two operational DMSP satellites that are positioned so that they cross the equator in the early morning, midmorning, and early afternoon. In addition, the government relies on a European satellite, called the Meteorological Operational (MetOp) satellite, for satellite observations in the midmorning orbit.[2] Together, the satellites ensure that, for any region of the earth, the data provided to users are generally no more than 6 hours old. Besides the operational satellites, six older satellites are in orbit that still collect some data and are available to provide limited backup to the operational satellites should they degrade or fail. The last POES satellite was launched in February 2009. The Air Force plans to launch its two remaining DMSP satellites as needed. Figure 1 illustrates the current operational polar satellite constellation.

---

[2]The European Organisation for the Exploitation of Meteorological Satellites' MetOp program is a series of three polar-orbiting satellites dedicated to operational meteorology. MetOp satellites are planned to be launched sequentially over 14 years. The first of these satellites was launched in 2006 and is currently operational. The next two are expected to launch in 2012 and 2017, respectively.

Figure 1: Configuration of Operational Polar Satellites

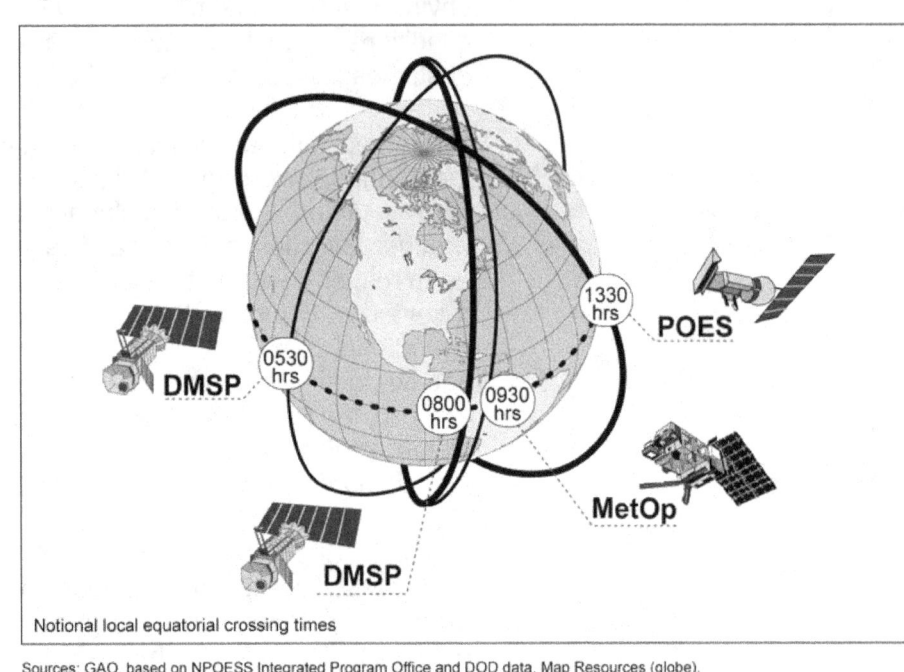

Notional local equatorial crossing times

Sources: GAO, based on NPOESS Integrated Program Office and DOD data, Map Resources (globe).

Polar Satellite Data and Products

Polar satellites gather a broad range of data that are transformed into a variety of products. Satellite sensors observe different bands of radiation wavelengths, called channels, which are used for remotely determining information about the earth's atmosphere, land surface, oceans, and the space environment. When first received, satellite data are considered raw data. To make them usable, processing centers format the data so that they are time-sequenced and include earth-location and calibration information. After formatting, these data are called raw data records. The centers further process these raw data records into channel-specific data sets, called sensor data records and temperature data records. These data records are then used to derive weather and climate products called environmental data records. These environmental data records include a wide range of atmospheric products detailing cloud coverage, temperature, humidity, and ozone distribution; land surface products showing snow cover, vegetation, and land use; ocean products depicting sea surface temperatures, sea ice, and wave height; and characterizations of the space environment. Combinations of these data records (raw, sensor, temperature, and environmental data records) are also used to derive more sophisticated products, including outputs from numerical weather models and assessments of climate trends. Figure 2 is a simplified depiction of the various stages of satellite data processing, and figure 3 depicts examples of two different weather products.

Figure 2: Stages of Satellite Data Processing

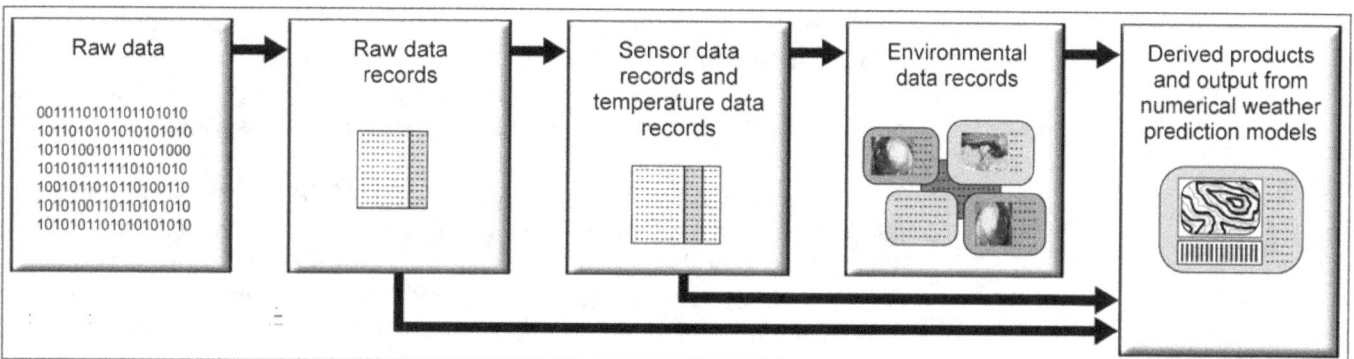

Source: GAO analysis of NOAA information.

Figure 3: Examples of Weather Products

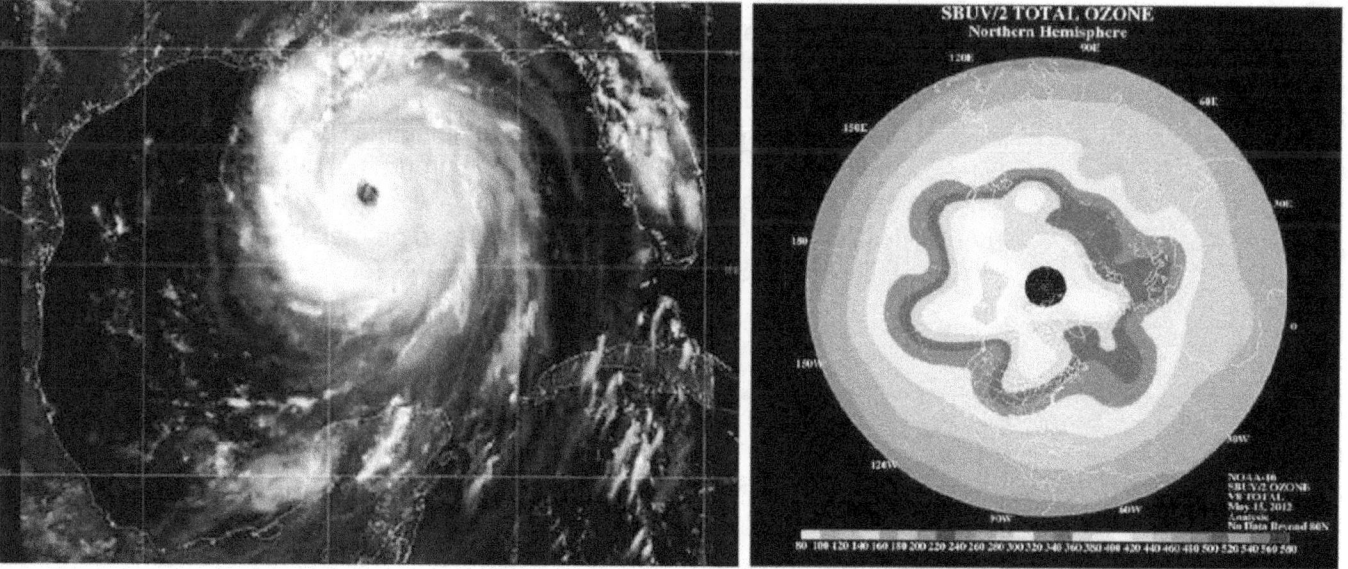

Source: NOAA's National Environmental Satellite Data and Information Service.

Note: The figure on the left is a POES Image of Hurricane Katrina in 2005; the figure on the right is an analysis of ozone concentration produced from POES satellite data.

GAO-12-604 Polar-Orbiting Environmental Satellites

The NPOESS Program: Inception, Challenges, and Divergence

With the expectation that combining the POES and DMSP programs would reduce duplication and result in sizable cost savings, a May 1994 Presidential Decision Directive required NOAA and DOD to converge the two satellite programs into a single satellite program—NPOESS—capable of satisfying both civilian and military requirements.[3] The converged program, NPOESS, was considered critical to the nation's ability to maintain the continuity of data required for weather forecasting and global climate monitoring. NPOESS satellites were expected to replace the POES and DMSP satellites in the morning, midmorning, and afternoon orbits when they neared the end of their expected life spans.

To manage this program, DOD, NOAA, and NASA formed a tri-agency Integrated Program Office, with NOAA responsible for overall program management for the converged system and for satellite operations, the Air Force responsible for acquisition, and NASA responsible for facilitating the development and incorporation of new technologies into the converged system.

When the primary NPOESS contract was awarded in August 2002, the program was estimated to cost about $7 billion through 2018. The program was to include the procurement and launch of 6 satellites over the life of the program, with each satellite hosting a subset of 13 instruments. The planned instruments included 11 environmental sensors, and two systems supporting specific user services (see table 1). To reduce the risk involved in developing new technologies and to maintain climate data continuity, the program planned to launch the NPP demonstration satellite in May 2006.[4] NPP was to demonstrate selected instruments that would later be included on the NPOESS satellites. The first NPOESS satellite was to be available for launch in March 2008.

---

[3]Presidential Decision Directive NSTC-2, May 5, 1994.

[4]In January 2012, the name of the satellite was changed to the Suomi National Polar-orbiting Partnership satellite. The NPP acronym remained the same.

Table 1: Anticipated NPOESS Instruments, as of July 2002

Instrument	Instrument type	Description
Advanced technology microwave sounder (ATMS)	Environmental sensor	Measures microwave energy released and scattered by the atmosphere; to be used in combination with the cross-track infrared sounder to produce daily global atmospheric temperature, humidity, and pressure profiles.
Aerosol polarimetry sensor	Environmental sensor	Retrieves specific aerosol (liquid droplets or solid particles suspended in the atmosphere, such as sea spray, smog, and smoke) and cloud measurements.
Conical microwave imager/sounder	Environmental sensor	Collects microwave images and data needed to measure rain rate, ocean surface wind speed and direction, amount of water in the clouds, and soil moisture, as well as temperature and humidity at different atmospheric levels.
Cross-track infrared sounder (CrIS)	Environmental sensor	Collects measurements of the infrared radiation emitted and scattered by the Earth and atmosphere to determine the vertical distribution of temperature, moisture, and pressure in the atmosphere.
Data collection system	System providing services to selected users	Collects environmental data from platforms around the world and delivers them to users worldwide.
Earth radiation budget sensor	Environmental sensor	Measures solar short-wave radiation and long-wave radiation released by the Earth back into space on a worldwide scale to enhance long-term climate studies.
Global positioning system occultation sensor	Environmental sensor	Measures the refraction of radio wave signals from the Global Positioning System and Russia's Global Navigation Satellite System to characterize the ionosphere and information related to the vertical distribution of temperature and moisture of the atmosphere.
Ozone mapper/ profiler suite (OMPS)	Environmental sensor	Collects data needed to measure the amount and distribution of ozone in the Earth's atmosphere. Consists of two components (nadir and limb) that can be provided separately.
Radar altimeter	Environmental sensor	Measures variances in sea surface height/topography and ocean surface roughness, which are used to determine sea surface height, significant wave height, and ocean surface wind speed and to provide critical inputs to ocean forecasting and climate prediction models.
Search and rescue satellite-aided tracking system	System providing services to selected users	A subsystem that detects and locates aviators, mariners, and land-based users in distress.
Space environmental sensor suite	Environmental sensor	Collects data to identify, reduce, and predict the effects of space weather on technological systems, including satellites and radio links.
Total and spectral solar irradiance sensor	Environmental sensor	Monitors and captures total and spectral solar irradiance data.
Visible/infrared imager radiometer suite (VIIRS)	Environmental sensor	Collects images and radiometric data used to provide information on the Earth's clouds, atmosphere, ocean, and land surfaces.

Source: GAO analysis of data from the former NPOESS Integrated Program Office.

In the years after the program was initiated, NPOESS encountered significant technical challenges in sensor development, program cost growth, and schedule delays. By November 2005, we estimated that the program's cost had grown to $10 billion, and the schedule for the first launch was delayed by almost 2 years.[5] These issues led to a 2006 decision to restructure the program, which reduced the program's functionality by decreasing the number of planned satellites from 6 to 4, and the number of instruments from 13 to 9. As part of the decision, officials decided to reduce the number of orbits from three (early morning, midmorning, and afternoon) to two (early morning and afternoon) and to rely solely on the European satellites for midmorning orbit data.

Even after the restructuring, however, the program continued to encounter technical issues in developing two sensors, significant tri-agency management challenges, schedule delays, and further cost increases. Because the schedule delays could lead to satellite data gaps, in March 2009, agency executives decided to use NPP as an operational satellite.[6] Later, in August 2009, faced with costs that were expected to reach about $15 billion and launch schedules that were delayed by over 5 years, the Executive Office of the President formed a task force, led by the Office of Science and Technology Policy, to investigate the management and acquisition options that would improve the NPOESS program. As a result of this review, in February 2010, the Director of the Office of Science and Technology Policy announced that NOAA and DOD would no longer jointly procure the NPOESS satellite system; instead, each agency would plan and acquire its own satellite system.[7] Specifically, NOAA would be responsible for the afternoon orbit and the observations planned for the first and third satellites. DOD would be responsible for the early morning orbit and the observations planned for the second and fourth satellites. The partnership with the European satellite agencies for the midmorning orbit was to continue as planned. When this decision was announced, NOAA immediately began planning for a new satellite program in the afternoon orbit—called JPSS—and DOD began planning for a new satellite program in the morning orbit—called DWSS.

---

[5]GAO, *Polar-Orbiting Operational Environmental Satellites: Technical Problems, Cost Increases, and Schedule Delays Trigger Need for Difficult Trade-off Decisions*, GAO-06-249T (Washington, D.C.: Nov. 16, 2005).

[6]Using NPP as an operational satellite means that the satellite's data will be used to provide climate and weather products.

[7]The announcement accompanied the release of the President's fiscal year 2011 budget request.

GAO-12-604 Polar-Orbiting Environmental Satellites

Overview of Initial NOAA and DOD Plans for Replacement Satellite Programs

After the decision was made to disband the NPOESS program in 2010, NOAA and DOD began planning for their respective satellite programs. For NOAA, these plans included:

- relying on NASA for system acquisition, engineering, and integration;

- completing, launching, and supporting NPP;

- acquiring and launching two satellites for the afternoon orbit, called JPSS-1 and JPSS-2;

- developing and integrating five sensors on the two satellites;

- finding alternate host satellites for selected instruments that would not be accommodated on the JPSS satellites; and

- providing ground system support for NPP, JPSS, and DWSS; data communications for MetOp and DMSP; and data processing for NOAA's use of microwave data from an international satellite.

In 2010, NOAA estimated that the life cycle costs of the JPSS program would be approximately $11.9 billion for a program lasting through fiscal year 2024, which included $2.9 billion in NOAA funds spent on NPOESS through fiscal year 2010.[8]

Alternatively, DOD planned that its DWSS program would be comprised of two satellites, the first to be launched no earlier than 2018. Each satellite was to have three sensors: a Visible/Infrared Imager/Radiometer Suite, a Space Environment Monitor, and a microwave imager/sounder. As of September 2011, DOD planned to conduct a thorough system requirements review before finalizing DWSS functionality, cost, and schedule. Table 2 compares the planned cost, schedule, and scope of the three satellite programs at different points in time.

---

[8]This figure does not include approximately $2.9 billion in sunk costs that DOD spent on NPOESS through fiscal year 2010.

GAO-12-604 Polar-Orbiting Environmental Satellites

Table 2: A Comparison of NPOESS and the NOAA and DOD Plans for Replacing NPOESS, at Different Points in Time

Key area	NPOESS program before it was restructured (as of May 2006)	NPOESS program after it was restructured (as of June 2006)	NPOESS program prior to being disbanded (as of February 2010)	NOAA and DOD acquisition plans (as of May 2010)
Life cycle range	1995-2020	1995-2026	1995-2026	JPSS: 1995-2024 DWSS: not determined
Estimated life cycle cost	$8.4 billion	$12.5 billion	$13.95+ billion[a]	JPSS: $11.9 billion (which includes about $2.9 billion spent through fiscal year 2010 on NPOESS) DWSS: not determined
Number of satellites	6 (in addition to NPP)	4 (in addition to NPP)	4 (in addition to NPP)	JPSS: 2 (in addition to NPP) DWSS: 2
Number of orbits	3 (early morning, midmorning, and afternoon)	2 (early morning and afternoon; would rely on European satellites for midmorning orbit data)	2 (early morning and afternoon; would rely on European satellites for midmorning orbit data)	JPSS: 1 (afternoon orbit) DWSS: 1 (early morning orbit) (European satellites would provide midmorning orbit)
Launch schedule	NPP by October 2006 First NPOESS (C1) by November 2009 Second NPOESS (C2) by June 2011	NPP by January 2010 C1 by January 2013 C2 by January 2016 C3 by January 2018 C4 by January 2020	NPP no earlier than September 2011 C1 by March 2014 C2 by May 2016 C3 by January 2018 C4 by January 2020	JPSS: • NPP –no earlier than September 2011 • JPSS-1 (C1 equivalent) available in 2015 • JPSS-2 (C3 equivalent) available in 2018 DWSS: no earlier than 2018
Number of sensors	11 sensors and 2 user services systems	NPP: 4 sensors C1: 6 sensors C2: 2 sensors C3: 6 sensors C4: 2 sensors	NPP: 5 sensors C1: 7 sensors[b] C2: 2 sensors C3: 6 sensors C4: 2 sensors	NPP: 5 sensors JPSS-1 and 2: 5 sensors[c] DWSS: 3 sensors

Source: GAO analysis of NOAA, DOD, and task force data.

[a]Although the program baseline was $13.95 billion in February 2010, we estimated in June 2009 that this cost could grow by about $1 billion. In addition, officials from the Executive Office of the President stated that they reviewed life-cycle cost estimates from DOD and the NPOESS program office of $15.1 billion and $16.45 billion, respectively.

[b]In May 2008, the NPOESS Executive Committee approved an additional sensor—the Total and Spectral Solar Irradiance Sensor—for the C1 satellite.

[c]The five sensors are ATMS, the Clouds and Earth's Radiant Energy System (CERES), CrIS, OMPS, and VIIRS. NOAA also committed to finding an alternative spacecraft and launch accommodation for the Total and Spectral Solar Irradiance Sensor, the Advanced Data Collection System, and the Search and Rescue Satellite-Aided Tracking system.

Prior GAO Work Recommended Actions to Solidify Plans and Address Risks

We have issued a series of reports on the NPOESS program highlighting technical issues, cost growth, and key management challenges affecting the tri-agency program structure.[9] For example, in June 2009, we added to our previous concerns about the tri-agency oversight of the NPOESS program.[10] We reported that the Executive Committee responsible for providing direction to the program was ineffective because the DOD acquisition executive did not attend committee meetings; the committee did not track action items to closure; and many of the committee's decisions did not achieve the desired outcomes. We also reported that the program's cost estimates were expected to rise and that the launch schedules were expected to be delayed. To help address these issues, we made recommendations to, among other things, improve executive-level oversight and develop realistic time frames for revising cost and schedule baselines. Agency officials agreed with our recommendations and took steps to improve executive oversight.

More recently, in May 2010, we reported on NOAA's and DOD's preliminary plans for initiating new environmental satellite programs and noted that agency officials had not yet made key decisions on their programs' cost, schedule, and capabilities and we highlighted key risks in transitioning from NPOESS to their new programs.[11] These risks included the loss of key staff and capabilities, delays in negotiating contract changes and establishing new program offices, the loss of support for the other agency's requirements, and insufficient oversight of new program management. We recommended that the Secretaries of Defense and Commerce direct their respective NPOESS follow-on programs to expedite decisions on the expected cost, schedule, and capabilities of their planned programs, and to develop plans to address the key transition risks we identified. As discussed later in this report, both agencies subsequently made decisions about the scope of their respective programs and took steps to mitigate the transition risks we

---

[9]For a full list of our reports on polar-orbiting environmental satellites, see "Related GAO Products" at the end of this report.

[10]GAO, *Polar-orbiting Environmental Satellites: With Costs Increasing and Data Continuity at Risk, Improvements Needed in Tri-agency Decision Making,* GAO-09-564 (Washington, D.C.: June 17, 2009).

[11]GAO, *Polar-Orbiting Environmental Satellites: Agencies Must Act Quickly to Address Risks That Jeopardize the Continuity of Weather and Climate Data,* GAO-10-558 (Washington, D.C.: May 27, 2010).

identified. For example, NOAA transferred key staff from the NPOESS program to the JPSS program and coordinated with the Air Force to negotiate contract changes.

Agencies Transferred Responsibilities to Their Respective Programs, but NOAA's Is Being Downsized, and DOD's Has Been Terminated

Following the decision to disband NPOESS, both NOAA and DOD were responsible for transferring key management responsibilities to their respective programs. This entailed (1) establishing separate program offices for their respective follow-on programs, (2) establishing requirements for their respective programs, and (3) transferring contracts from NPOESS to the new programs.

Both agencies made progress on these activities, but recent events have resulted in major program changes. Specifically, NOAA established its JPSS program office, established program requirements, and transferred most sensor contracts. However, the agency now plans to remove key requirements, including selected sensors and ground systems, to keep the program within budget. DOD established its DWSS program office and modified its contracts accordingly before deciding in early 2012 to terminate the program and reassess its requirements (as directed by Congress).

NOAA Established the JPSS Program and Contracts for Most Components, but Plans to Modify Requirements to Limit Costs

After the February 2010 decision to disband NPOESS, NOAA transferred management responsibilities to its new satellite program, defined its requirements, and transferred contracts to the new program. Specifically, NOAA established a program office to guide the development of the NPP and JPSS satellites. NOAA also worked with NASA to establish its program office to oversee the acquisition, system engineering, and integration of the satellite program. By 2011, the two agencies had established separate—but colocated—JPSS program offices, each with different roles and responsibilities. NOAA's program office is responsible for programmatic activities related to the satellites' development, including managing requirements, budgets, and interactions with satellite data users. Alternatively, NASA's program office is responsible for the development and integration of the sensors, satellites, and ground systems. In January 2012, both agencies approved a management control plan that delineates the two agencies' roles, responsibilities, and executive oversight structure.

In September 2011, NOAA established its official requirements document for the JPSS program. This document defines the components of the program as well as the expected performance of the satellites and ground systems. Key components include NPP, the two JPSS satellites, the five sensors, a distributed ground-based network of satellite data receptor sites, and four ground-based data processing systems. This system is to deliver 31 satellite data products within 80 minutes of observation on the first satellite and within 30 minutes on the second satellite.

Over the 2 years since the decision to disband NPOESS, NOAA has also been working to transfer and refine the contracts for four of the sensors that are to be launched on the first JPSS satellite from the Air Force to NASA.[12] The program completed the transfer of all of the contracts by September 2011 and then began the process of updating the contracts to match JPSS' requirements. This process has been completed for three sensors (CrIS, OMPS, and ATMS). Program officials expect to finalize changes to the contract for the last sensor (VIIRS) in June 2012.

NOAA Plans to Modify Requirements and May Remove Program Elements to Keep Costs at $12.9 Billion

While NOAA and NASA have made progress in transferring management and contract responsibilities from NPOESS to the JPSS program, NOAA recently decided to modify its requirements in order to limit program costs. From January to December 2011, the agency went through a cost estimating exercise for the JPSS program. This exercise included identifying key program elements, documenting assumptions, performing historical and parametric analysis to determine reasonable estimates for the elements, seeking an independent cost estimate, and reconciling the two estimates. At the end of this exercise, NOAA validated that the cost of the full set of JPSS functions from fiscal year 2012 through fiscal year 2028 would be $11.3 billion. After adding the agency's sunk costs of $3.3 billion, the program's life cycle cost estimate totaled $14.6 billion.[13] This amount is $2.7 billion higher than the $11.9 billion estimate for JPSS when NPOESS was disbanded in 2010. According to NOAA officials, this increase is primarily due to a 4-year extension of the program from 2024 to 2028, the addition of previously unbudgeted items such as the free flyers, cost growth associated with transitioning contracts from DOD to NOAA, and the program's decision to slow down work on lower-priority elements because of budget constraints in 2011.

---

[12]CERES was already a NASA contract during the NPOESS program, so this contract did not need to be transferred from the Air Force.

[13]NOAA's $3.3 billion sunk costs included $2.9 billion through fiscal year 2010 and about $400 million in fiscal year 2011.

In working with the Office of Management and Budget to establish the president's fiscal year 2013 budget request, NOAA officials stated that they agreed to fund JPSS at roughly $900 million per year through 2017, to merge funding for two climate sensors into the JPSS budget, and to cap the JPSS life cycle cost at $12.9 billion through 2028. Because this cap is $1.7 billion below the expected $14.6 billion life cycle cost of the full program, NOAA decided to remove selected elements from the satellite program. While final decisions on what will be removed are expected by the end of June 2012, NOAA may discontinue:

- support for OMPS operations on JPSS-1;

- development of two of the three planned Total and Spectral Solar Irradiance Sensors, the spacecraft for all three of these sensors, and the launch vehicle for the three sensors;

- development of the OMPS and CERES sensors on JPSS-2;

- plans for a network of ground-based receptor stations;

- planned improvements in the time it takes to obtain satellite data from JPSS-2 (the requirement was to provide data in 30 minutes; instead, the requirement will remain at the JPSS-1 level of 80 minutes);

- plans to install an Interface Data Processing Segment (IDPS) at two Navy locations; and

- plans to support ground operations for DOD's future polar satellite program.

NOAA anticipates modifying its official requirements documents to reflect these changes by the end of 2012. The removal of these elements will affect both civilian and military satellite data users. The loss of OMPS and CERES satellite data could cause a break in the over 30-year history of satellite data and would hinder the efforts of climatologists and meteorologists focusing on understanding changes in the earth's ozone coverage and radiation budget.[14] The loss of ground-based receptor stations means that NOAA may not be able to improve the timeliness of

---

[14]The radiation budget is the amount of the solar energy entering and leaving the earth's atmosphere.

GAO-12-604 Polar-Orbiting Environmental Satellites

JPSS-2 satellite data from 80 minutes to the current 30 minute requirement, and as a result, weather forecasters will not be able to update their weather models using the most recent satellite observations. Further, the loss of the data processing systems at the two Navy locations means that NOAA and the Navy will need to establish an alternative way to provide data to the Navy.

DOD Established and Subsequently Terminated Its DWSS Program

After the February 2010 decision to disband NPOESS, DOD transferred management responsibilities to its new satellite program, started defining its requirements, and modified contracts to reflect the new program. Specifically, in 2010, DOD established a DWSS program office and started developing plans for what the satellite program would entail. The DWSS program office, located at the Space and Missile Systems Center in Los Angeles, California, was given responsibility for the acquisition, development, integration, and launch of the DWSS satellites. Because this is considered a major acquisition, it is overseen by the Defense Acquisition Board and the Under Secretary of Defense for Acquisition, Technology, and Logistics.

In August 2010, the agency determined that the DWSS program would include two satellites and that each satellite would host three sensors.[15] Over the following year, the program office developed a program plan and a technical description, and planned to define requirements in early 2012. Further, the agency started modifying its existing contracts with the NPOESS contractor to reflect the new program. By May 2011, the program office had contracted for DWSS activities through the end of 2012.

These efforts, however, have been halted. In early 2012, with congressional direction, DOD decided to terminate the DWSS program because it still has two DMSP satellites to launch and it did not yet need the DWSS satellites.[16] In January 2012, the Air Force halted work on the

---

[15]These sensors included VIIRS, a space environment monitor, and a microwave imager/sounder.

[16]DOD officials stated that they plan to launch DMSP-19 in 2014 and DMSP-20 when needed. If DMSP-19 lasts 6 years, there is a chance that DMSP-20 would not be launched until 2020. Thus, in a best-case scenario, the follow-on satellites would not need to be launched until 2026.

program. DOD is currently identifying alternative means to fulfill its future environmental satellite requirements.

NPP Is in Orbit and Transmitting Data; Development of the First JPSS Satellite Has Begun, but Critical Steps Remain

In September 2010, shortly after NPOESS was disbanded, NOAA and NASA established plans for both NPP and JPSS. These plans included

- launching NPP by the end of October 2011 and completing an early on-orbit check out of the NPP spacecraft and sensors (called commissioning) by the end of January 2012;

- completing all NPP calibration and validation activities[17] by October 2013; and

- developing, testing, and launching JPSS-1 by the end of 2014 and JPSS-2 by the end of 2017.[18]

Program officials currently estimate that JPSS-1 will launch by March 2017 and JPSS-2 will launch by December 2022. NOAA officials explained that part of the reason for the change in launch dates is that the program's budget under the 2011 continuing resolution was only one third of what NOAA had anticipated. Thus, program officials decided to defer development of the first JPSS satellite in order to keep NPP on track.

---

[17]After a satellite has been launched, scientists perform an on-orbit accuracy check, called calibration and validation, to verify that the sensors accurately report ground and atmospheric conditions. These activities ensure that satellite data products are ready for operational use.

[18]NOAA officials noted that the JPSS launch dates could change as the agency finalized its program planning activities.

NPP Is in Orbit; Sensor Data Are Being Calibrated for Use

NPP was successfully launched on October 28, 2011. After launch, NASA began the process of activating the satellite and commissioning the instruments. This process ended at the beginning of March 2012, which was a little over a month after the planned completion date at the end of January 2012. The delay was caused by an issue on the VIIRS instrument that caused the program to halt commissioning activities in order to diagnose the problem. Specifically, the quality of VIIRS data in certain bands was degrading much more quickly than expected. NASA and the JPSS program office subsequently identified the problem as contamination on VIIRS mirrors. NOAA and NASA program officials, including the JPSS director and project manager, reported that this issue is not expected to cause the instrument to fall below its performance specifications. Figure 4 depicts an image of Earth using VIIRS data from NPP.

Figure 4: Composite of VIIRS Images

Source: NASA.

Program officials are working to complete NPP calibration and validation activities by October 2013, but they acknowledge that they may encounter delays in developing satellite products. NOAA is receiving data from the five sensors on the NPP satellite, and has begun calibration and validation. According to NOAA and NASA officials, during this time, the products go through various levels of validation, including a beta stage (products have been minimally validated, but are available to users so that they can begin working with the data); a provisional stage (products are not optimal, but are ready for operational evaluation by users); and a validated stage (products are ready for operational use). The amount of time it takes for a product to be fully validated depends on the sensor and the type of product. For example, NOAA provided a provisional ozone environmental data record from the OMPS sensor in April 2012 and expects to provide three beta environmental data records from the CrIS sensor by October 2012. NOAA's users began to use validated ATMS products in May 2012, and NOAA expects that they will increase the amount and types of data they use in the following months.

Development of JPSS Is Under Way; Critical Decisions and Milestones Are Pending

The major components of the JPSS program are at different stages of development, and important decisions and program milestones lie ahead. NASA's JPSS program office organized its responsibilities into three separate projects: (1) the flight project, which includes sensors, spacecraft, and launch vehicles; (2) the ground project, which includes ground-based data processing and command and control systems, and (3) the free-flyer project, which involves developing and launching the instruments that are not going to be included on the JPSS satellites. Table 3 shows the three JPSS projects and their key components.

Table 3: JPSS Projects and Components

Project	Key components and responsibilities
Flight	• Sensors: ATMS, CERES, CrIS, OMPS, VIIRS
	• Spacecraft
	• Launch vehicle
Ground	• Satellite command, control, and communications
	• Interface data processing segment (IDPS)
Free Flyer	• Sensor: Total and Spectral Solar Irradiance Sensor
	• Subsystems: Search and Rescue Satellite-aided Tracking System and the Advanced Data Collection System
	• Spacecraft
	• Launch vehicle

Source: NOAA's JPSS Program Office.

Within the flight project, development of the sensors for the first JPSS satellite is well under way; however, selected sensors are experiencing technical issues and the impact of these issues had not yet been determined. The ground project is currently in operation supporting NPP, and NOAA is planning to upgrade selected parts of the ground systems to increase security and reliability. The free-flyer project is still in a planning stage because NOAA has not yet decided which satellites will host the instruments or when these satellites will launch. One of these projects has recently completed a major milestone and one project has its next milestone approaching. Specifically, the flight project completed a separate system requirements review in April 2012, while the ground project's system requirements review is scheduled for August 2012.

Flight Project: Development of JPSS-1 Sensors Is Well Under Way, but Technical Issues Continue to Surface

Because development of the sensors for JPSS-1 began during the NPOESS era, NASA estimates that as of March 2012, all of the sensors have been at least 60 percent completed. However, selected sensors are encountering technical issues and the full impact of these issues on cost and schedule has not been determined. Further, the program has not yet made a decision on which launch vehicle will be used. NASA and NOAA officials reported that the technical issues thus far are routine in nature, and that they plan to select a launch vehicle by the end of 2012. Table 4 describes the current status of the components of the JPSS-1 flight project.

Table 4: Status of Key Components of the Flight Project Supporting the First JPSS Satellite, as of April 2012

Component	Status
Sensor	
ATMS	NASA estimates that this sensor is about 85 percent complete. However, the sensor is experiencing technical issues in development. For example, a mixer used in support of high-frequency sensing failed during development, and needs to be replaced. Program officials have not yet determined the impact of this issue on the sensor's cost and schedule.
CERES	According to NASA, this sensor has been developed, and completed environmental testing in May 2012. Pre-ship review is currently planned for June 2012.
CrIS	NASA estimates that this sensor is about 60 percent complete. The program office decided to move to a new data bus[a] because the one used on NPP is now obsolete. At present, this sensor has an 8-month schedule buffer.
OMPS- Nadir	NASA estimates that OMPS-Nadir is approximately 70 percent complete. Unlike NPP, there will not be a Limb component on JPSS-1.
VIIRS	NASA estimates the sensor is about 65 percent complete. The sensor has encountered a number of technical and workmanship issues that need to be addressed. For example, program and contract officials reported experiencing problems with the manufacturing and testing of parts by subcontractors, including a recent issue with one of the sensor's mirrors. In addition, the program office decided to move to a new data bus because the one used on NPP is now obsolete. Further, the program will need to address communication issues that could affect VIIRS' ability to provide data in every orbit, but they have not identified the potential cost and schedule impact of this issue.
Spacecraft	Development of the spacecraft has begun. A key milestone, the spacecraft's critical design review, is currently planned for early 2013. The spacecraft is currently on the critical path, which means that any delay in development will delay the launch date of the satellite.
Launch vehicle	The launch vehicle is currently being competed and NASA officials anticipate making a selection this year. However, each potential launch vehicle has different risks and costs associated with it. We recently reported that multiple NASA projects have encountered challenges with their launch vehicles, including the cost and availability of the vehicles and the uncertainty of using new, unproven vehicles.[b]

Source: GAO analysis of NOAA and NASA data.

[a]A data bus is used to allow high-speed communications and data transfer between the sensor and the spacecraft.

[b]GAO, *NASA: Assessments of Selected Large-Scale Programs*, GAO-12-207SP (Washington, D.C.: Mar. 1, 2012).

Ground Project: NOAA Plans Upgrades to Strengthen the Infrastructure

While NOAA ground systems for satellite command, control, and communications and for data processing are currently supporting NPP operations, the agency plans to upgrade the ground systems to improve their availability and reliability. In 2010, we reported that NPP's ground systems had weaknesses because they were developed using outdated security requirements approved in 1998. These weaknesses were highlighted soon after NPP was launched, when the communications links providing satellite data from the satellite receiver in Svalbard, Norway, to the United States were severed. NOAA immediately established a temporary backup capability, and plans to upgrade its communications systems to establish permanent backup capabilities by the end of 2012. In addition, NOAA plans to enhance the backup capabilities of its data processing system infrastructure by November 2015.

Free Flyer Project: Key Decisions Have Been Made; Others Are Pending

The instruments in the free flyer project, including the Total and Spectral Solar Irradiance Sensor and two user services systems (the Search and Rescue Satellite-Aided Tracking system and an Advanced Data Collection system), are currently under development. However, in early 2012, NOAA decided to consider not launching the Total and Spectral Solar Irradiance Sensor as an option for staying within its budget cap. Moreover, the agency is still considering its options for the spacecraft that will carry the other two instruments to space. For example, it is considering contracting for a spacecraft or having the instruments hosted on some other organization's satellite. Table 5 depicts the status of the components of the free-flyer project.

Table 5: Status of Key Free Flyer Components, as of April 2012

Component	Status
Total and Spectral Solar Irradiance Sensor	Delivery of the first of three sensors is expected in March 2013.
Advanced Data Collection System	Delivery of this system by France's space agency is expected in May 2013.
Search and Rescue Satellite-Aided Tracking system	While one component of this system is five months behind schedule, delivery of this system, provided jointly by the France and Canadian space agencies, is scheduled for January 2015.
Free Flyer-1 spacecraft	NOAA is considering the possibility of not acquiring a spacecraft for the Total and Spectral Solar Irradiance Sensor. It is considering its alternatives for a spacecraft for the two other subsystems.
Launch vehicle	NOAA is considering the possibility of not launching the Total and Spectral Solar Irradiance Sensor. The program office has not decided on the launch vehicle for the two other subsystems.

Source: GAO analysis of NOAA and NASA data.

JPSS Risk Management Process in Place; Key Risks Remain

The JPSS program has a structured risk management process in place and is working to mitigate key program risks; however, NOAA faces key risks involving the potential for satellite gaps and does not yet have mitigation plans. According to best practices advocated by leading system engineering and program management organizations, effective risk management addresses four key areas: preparing for risk management, identifying and analyzing risks, mitigating risks, and providing executive oversight.[19] The JPSS program office has implemented elements of an effective risk management process. Specifically, the program documented its risk management strategy; identified relevant stakeholders and designated responsibilities for risk management activities; established and implemented standards for categorizing and prioritizing risks; instituted a program to identify, track, and mitigate risks; and established a process for regularly communicating risks to senior NASA and NOAA management.

The JPSS program is working to mitigate the risks of a lack of a cost and schedule baseline and program office staffing shortfalls, but NOAA has not established mitigation plans to address the risk of a gap in the afternoon orbit or potential satellite data gaps in the DOD and European polar satellite programs, which provide supplementary information to NOAA forecasts. Because it could take time to adapt grounds systems to receive alternative satellites' data, delays in establishing mitigation plans could leave the agency little time to leverage its alternatives. Until NOAA identifies its mitigation options, it may miss opportunities to leverage alternative satellite data sources. Moreover, until NOAA establishes mitigation plans for a satellite data gap, it runs the risk of not being able to fulfill its mission of providing weather forecasts to protect lives, property, and commerce.

NOAA Is Working to Mitigate Delays in Establishing Cost and Schedule Baselines

NOAA oversaw the establishment of contracts for the JPSS-1 sensors and spacecraft and NASA is managing the cost, schedule, and deliverables on these contracts using discrete task orders, but the agencies have not established a contractual cost and schedule baseline that would allow them to monitor contractor deliverables within an earned value management system.[20] In addition, program officials have not yet established an overall program baseline that delineates the cost, schedule, and content of the entire program. Under NASA's acquisition life cycle, a program baseline is due at the key decision milestone

---

[19]See, for example, Carnegie Mellon University's Software Engineering Institute, *Capability Maturity Model® Integration for Acquisition, Version 1.3 (CMMI-ACQ, V1.3)* and Project Management Institute Inc., *A Guide to the Project Management Body of Knowledge (PMBOK® Guide) – Third Edition*, (Newtown Square, PA: 2004).

[20]Earned value management is a technique for monitoring the cost and schedule of each work activity performed. It allows managers insights into the contractor's productivity and provides an early warning on cost and schedule variances.

scheduled to be completed by July 2013. Managing a program without a baseline makes it more difficult for program officials to make informed decisions and for program overseers to understand if the program is on track to successfully deliver expected functionality on cost and schedule.

Program officials acknowledge that the lack of a baseline is a risk, and they are tracking it through their risk management program. Program officials explained that after transferring the contracts from the Air Force to NASA, they needed to definitize the contracts to reflect JPSS program requirements instead of NPOESS program requirements. The JPSS program office has completed this process for three sensors (CrIS, OMPS, and ATMS) and is working to complete the process for one other sensor (VIIRS) by June 2012. After definitizing each contract to JPSS requirements and schedules, NASA and the contractors will perform an integrated baseline review before implementing an earned value management system. NOAA officials reported that they are working to establish contractual baselines as rapidly as practical for each of the contracts.

Program officials also plan to establish an overall program baseline. Actions planned to mitigate this risk include

- establishing a stable and realistic 5-year budget profile, which was completed in December 2011;

- refining the program requirements to match the expected budget by October 2012;

- definitizing contracts to address any changes in requirements in September 2012; and

- establishing the overall program baseline by the end of November 2012.

NOAA Is Working to Mitigate Risks in Program Staffing

NOAA and NASA have not yet fully staffed their respective JPSS program offices. While having a knowledgeable and capable program management staff is essential to any acquisition program, it is especially critical given the history of management challenges on the NPOESS program. However, NOAA has not yet filled 18 of the 64 positions it plans for the program office, including those for a program scientist and system engineers for the JPSS satellite, ground systems, and overall mission. In addition, NASA has not yet filled 6 positions it plans for its ground project.

Until these positions are filled, other staff members are supporting the workload and this could delay the schedule for implementing improvements in the ground systems.

Both agencies are actively tracking their respective program offices' staffing and plans for filling vacancies. According to NOAA officials, the agency is mitigating this risk by filling three of the vacant positions with long-term detailees. Further, NOAA plans to fill most of the positions, including that of the technical director, by July 2012. NASA has started the process to fill its vacancies, and plans to fill these by the end of September 2012.

NOAA Has Not Established Plans to Mitigate an Expected Gap in Satellite Data Continuity

In September 2011, we reported that NOAA was facing a gap in satellite data continuity; the risk of that gap is higher today. When NPOESS was first disbanded, program officials anticipated launching the JPSS satellites in 2015 and 2018 (while acknowledging that these dates could change as the program's plans were firmed up). Over the past year, as program officials made critical decisions to defer work on JPSS in order to keep NPP on track, the launch dates for JPSS-1 and JPSS-2 have changed. Program officials currently estimate that JPSS-1 will be launched by March 2017 and JPSS-2 will be launched by December 2022.

NOAA officials acknowledge that there is a substantial risk of a gap in satellite data in the afternoon orbit, between the time when the NPP satellite is expected to reach the end of its life and the time when the JPSS-1 satellite is to be in orbit and operational. This gap could span from 17 months to 3 years or more. In one scenario, NPP would last its full expected 5-year life (to October 2016), and JPSS-1 would launch as soon as possible (in March 2017) and undergo on-orbit checkout for a year (until March 2018). In that case, the data gap would extend 17 months. In another scenario, NPP would last only 3 years as noted by NASA managers concerned with the workmanship of selected NPP sensors. Assuming that the JPSS-1 launch occurred in March 2017 and the satellite data was certified for official use by March 2018, this gap would extend for 41 months. Of course, any problems with JPSS-1 development could delay the launch date and extend the gap period. Given the history of technical issues and delays in the development of the NPP sensors and the current technical issues on the sensors, it is likely that the launch of JPSS-1 will be delayed. Figure 5 depicts four possible gap scenarios.

Figure 5: Potential Gaps in Polar Satellite Data in the Afternoon Orbit

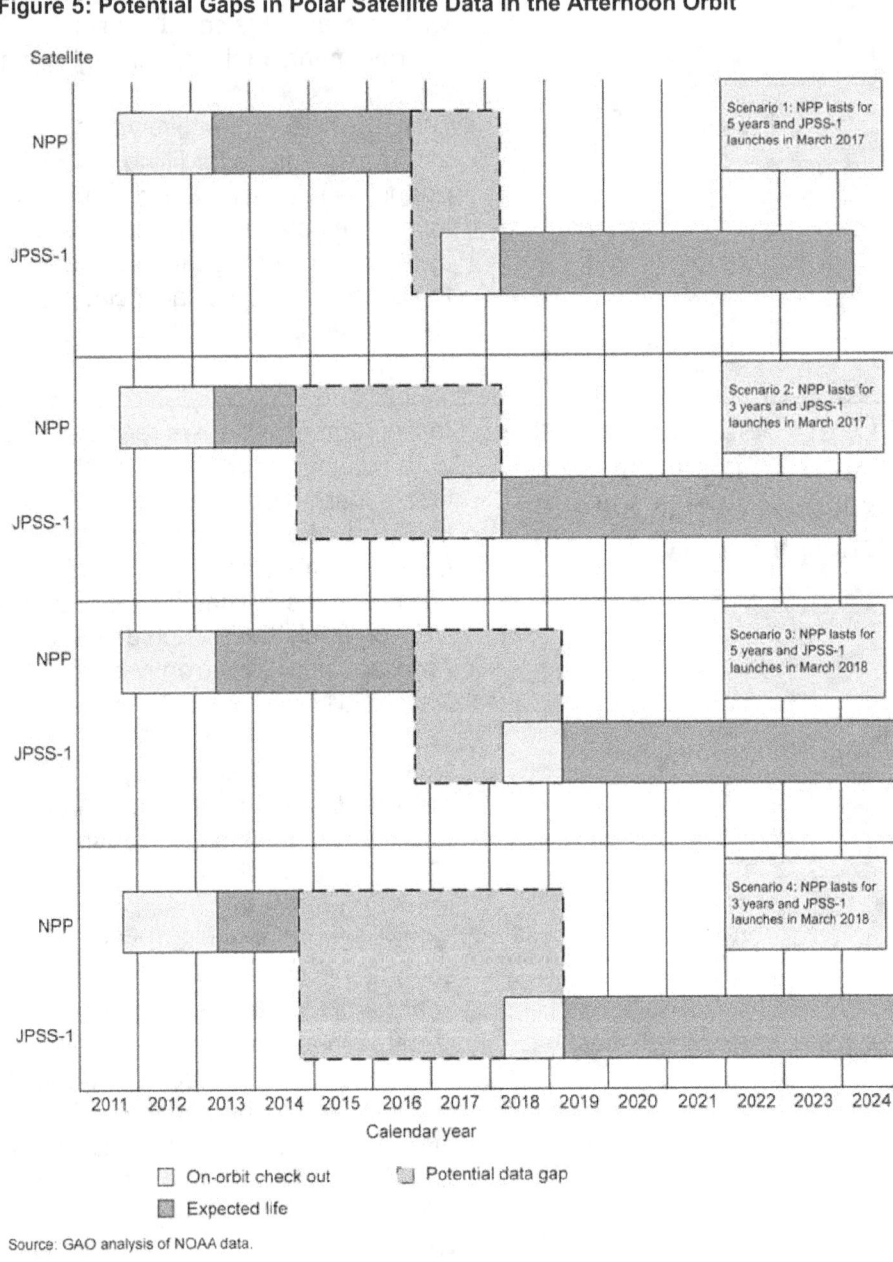

Source: GAO analysis of NOAA data.

According to NOAA, a data gap would lead to less accurate and timely weather prediction models used to support weather forecasting, and advanced warning of extreme events—such as hurricanes, storm surges, and floods—would be diminished. To illustrate this, the National Weather

Service performed several case studies to demonstrate how its weather forecasts would have been affected if there were no polar satellite data in the afternoon orbit. For example, when the polar satellite data were not used to predict the "Snowmaggedon" winter storm that hit the Mid-Atlantic coast in February 2010, weather forecasts predicted a less intense storm, slightly further east, and producing half of the precipitation at 3, 4, and 5 days before the event. Specifically, weather prediction models under-forecasted the amount of snow by at least 10 inches. The agency noted that this level of degradation in weather forecasts could place lives, property, and critical infrastructure in danger.

NOAA officials have communicated publicly and often about the risk of a satellite data gap; however, the agency has not established plans to mitigate the gap. NOAA officials stated that the agency will continue to use existing POES satellites, as well as NPP, as long as they provide data and that there are no viable alternatives to the JPSS program. However, it is possible that other governmental, commercial, or international satellites could supplement the data. If there are viable options for obtaining data from external sources, it would take time to adapt NOAA systems to receive, process, and disseminate the data. Until NOAA identifies these options, it may miss opportunities to leverage these satellite data sources.

NOAA Has Not Established Plans to Mitigate the Risk That the Polar Satellite Constellation Is Becoming Increasingly Unreliable

Since its inception, NPOESS was seen as a constellation of satellites providing observations in the early morning, midmorning, and afternoon orbits. Having satellites in each of these orbits ensures that satellite observations covering the entire globe are no more than 6 hours old, thereby allowing for more accurate weather predictions. Even after the program was restructured in 2006 and eventually terminated in 2010, program officials and the administration planned to ensure coverage in the early morning, midmorning, and afternoon orbits by relying on DOD satellites for the early morning orbit, the European satellite program for the midmorning, and NOAA's JPSS program for the afternoon orbit. However, recent events have made the future of this constellation uncertain:

- **Early morning orbit**—As discussed earlier in this report, in early fiscal year 2012, DOD terminated its DWSS program. While the agency has two more satellites to launch and is working to develop alternative plans for a follow-on satellite program, there are considerable challenges in ensuring that a new program is in place and integrated with existing ground systems and data networks in time to avoid a gap in this orbit.

DOD officials stated that they plan to launch DMSP-19 in 2014 and DMSP-20 when it is needed. If DMSP-19 lasts 6 years, there is a chance that DMSP will not be launched until 2020. Thus, in a best-case scenario, the follow-on satellites will not need to be launched until roughly 2026. However, civilian and military satellite experts have expressed concern that the DMSP satellites are quite old and may not work as intended. If they do not perform well, DOD could be facing a satellite data gap in the early morning orbit as early as 2014.

- **Midmorning orbit**—The European satellite organization plans to continue to launch MetOp satellites that will provide observations in the midmorning orbit through October 2021. The organization is also working to define and gain support for the follow-on program, called the Eumetsat Polar System-2nd Generation program. However, in 2011, NOAA alerted European officials that, because of the constrained budgetary environment, they will no longer be able to provide sensors for the follow-on program. Due to the uncertainty surrounding the program, there is a chance that the first European follow-on satellite will not be ready in time to replace MetOp at the end of its expected life. In that case, this orbit, too, would be in jeopardy.

- **Afternoon orbit**—As discussed previously, there is likely to be a gap in satellite observations in the afternoon orbit that could last well over one year. While our scenarios demonstrated gaps lasting between 17 and 53 months, NOAA program officials believe that the most likely scenario involves a gap lasting 18 to 24 months.

Figure 6 depicts the polar satellite constellation and the uncertain future coverage in selected orbits.

Figure 6: The Polar Satellite Constellation

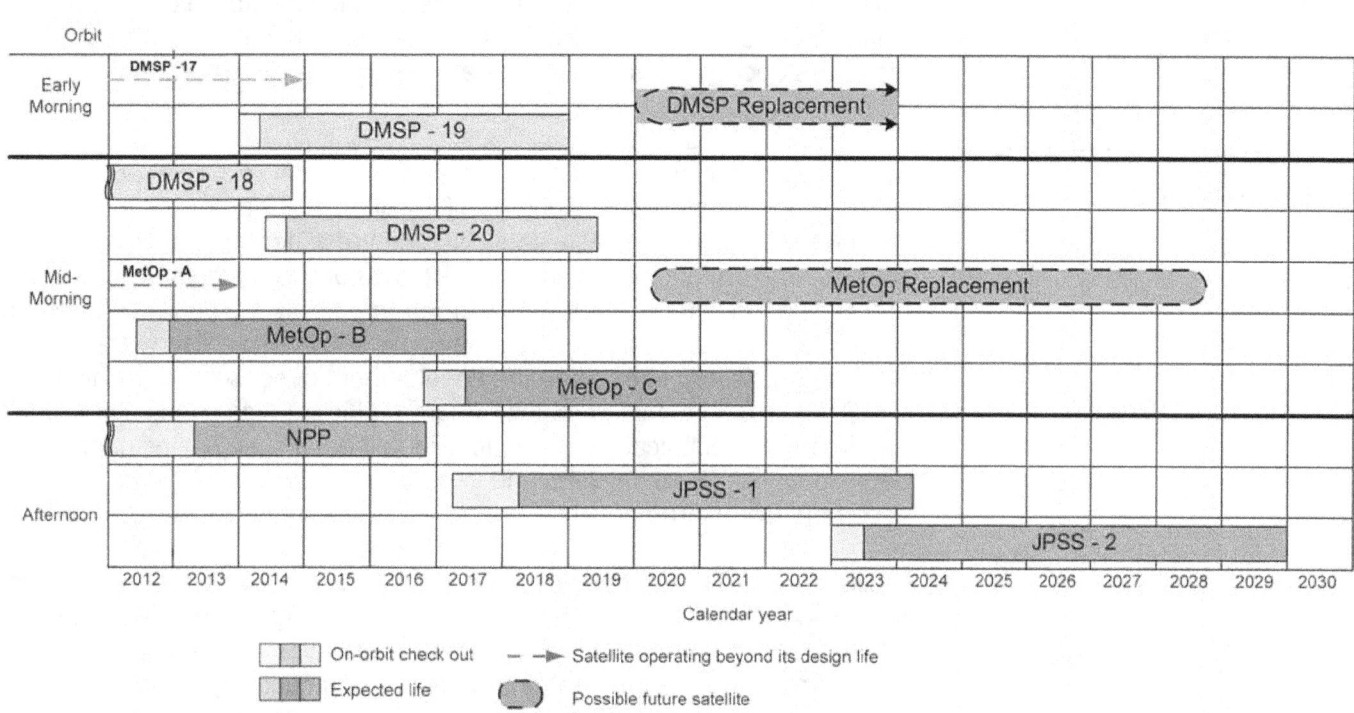

The NOAA Administrator and other senior executives acknowledge the risk of a data gap in each of the orbits of the polar satellite constellation and are working with European and DOD counterparts to coordinate their respective requirements and plans; however, they have not established plans for mitigating risks to the polar satellite constellation. As in the case of the anticipated gap in the afternoon orbit, NOAA plans to use older polar satellites to provide some of the necessary data for the other orbits. However, it is also possible that other governmental, commercial, or international satellites could supplement the data. For example, foreign nations continue to launch polar-orbiting weather satellites to acquire data such as sea surface temperatures, sea surface winds, and water vapor. Also, over the next few years, NASA plans to launch satellites that will collect information on precipitation and soil moisture.[21] If there are viable

---

[21]NASA plans to launch the Global Precipitation Measurement Mission satellite by June 2014 and the Soil Moisture Active and Passive satellite by January 2015.

GAO-12-604 Polar-Orbiting Environmental Satellites

options from external sources, it could take time to adapt NOAA systems to receive, process, and disseminate the data to its satellite data users. Until NOAA identifies these options and establishes mitigation plans, it may miss opportunities to leverage alternative satellite data sources.

Conclusions

After spending about $3.3 billion on the now-defunct NPOESS program, NOAA officials have established a $12.9-billion JPSS program and made progress in launching NPP, establishing contracts for the first JPSS satellite, and enhancing the ground systems controlling the satellites and processing the satellite data. JPSS program officials are currently working to calibrate NPP data so that they are useable by civilian and military meteorologists and to manage the development of sensors for the first JPSS satellite. In coming months, program officials face changing requirements, technical issues on individual sensors, key milestones in developing the JPSS satellite, and important decisions on how to accommodate instruments that are not included on the JPSS satellite.

While the JPSS program office is working to mitigate risks associated with not having a program baseline or a fully staffed program management office, NOAA has not established plans to mitigate the almost certain satellite data gaps in the afternoon orbit or the potential gaps in the early and mid-morning orbits. These gaps will likely affect the accuracy and timeliness of weather predictions and forecasts and could affect lives, property, military operations, and commerce. Because it could take time to adapt ground systems to receive an alternative satellite's data, delays in establishing mitigation plans could leave the agency little time to leverage alternatives. Until NOAA identifies its mitigation options, it may miss opportunities to leverage alternative satellite data sources.

Recommendations for Executive Action

Given the importance of polar-orbiting satellite data to weather forecasts, we recommend that the Secretary of Commerce direct the Administrator of NOAA to establish mitigation plans for risks associated with pending satellite data gaps in the afternoon orbit as well as potential gaps in the early morning and midmorning orbits.

Agency Comments and Our Evaluation

We sought comments on a draft of our report from the Department of Commerce, DOD, and NASA. We received written comments from the Secretary of Commerce, who transmitted NOAA's comments. In its comments, NOAA agreed with the report's recommendation and noted that the National Environmental Satellite, Data, and Information Service—

a NOAA component agency—has performed analyses on how to mitigate potential gaps in satellite data, but has not yet compiled this information into a report. The agency plans to provide a report to NOAA by August 2012. The department's comments are provided in appendix II. The department also provided technical comments, which we incorporated as appropriate.

While neither DOD nor NASA provided comments on the report's findings or recommendations, they offered technical comments, which we incorporated as appropriate. Specifically, the Staff Action Officer for the Space and Intelligence Office within the Office of the Under Secretary of Defense for Acquisition, Technology, and Logistics provided technical comments both orally and via e-mail, and a commander within the Navy's Oceanographer staff provided oral technical comments. In addition, the Project Manager of the JPSS flight project—a NASA employee—provided technical comments via e-mail.

As agreed with your office, unless you publicly announce the contents of this report earlier, we plan no further distribution of it until 30 days from the date of this letter. We are sending copies of this report to interested congressional committees, the Secretary of Commerce, the Secretary of Defense, the Administrator of NASA, the Director of the Office of Management and Budget, and other interested parties. In addition, this report will be available on the GAO Web site at http://www.gao.gov.

If you have any questions about this report, please contact me at (202) 512-9286 or at pownerd@gao.gov. Contact points for our Offices of Congressional Relations and Public Affairs may be found on the last page of this report. GAO staff who made major contributions to this report are listed in appendix III.

David A. Powner
Director, Information Technology Management Issues

Appendix I: Objectives, Scope, and Methodology

Our objectives were to (1) evaluate efforts to transfer management and contract responsibilities from the National Polar-orbiting Operational Environmental Satellite System (NPOESS) program to the separate satellite programs being established at the National Oceanic and Atmospheric Administration (NOAA) and Department of Defense (DOD), (2) assess NOAA's progress in developing the NPOESS Preparatory Project (NPP) satellite and the Joint Polar Satellite System (JPSS), and (3) evaluate NOAA's efforts to mitigate key project risks.

To evaluate efforts to transfer responsibilities from NPOESS to the separate NOAA and DOD programs, we compared the agencies' plans for establishing program management offices, developing program requirements, and transferring contracts to each agency's actual accomplishments. We analyzed key program documents, including acquisition decision memorandums, requirements documents, and the management control plan. We observed NOAA's monthly program management briefings and obtained detailed briefings on efforts to establish a program cost estimate, NOAA's fiscal year 2013 budget for JPSS, and decisions to remove selected program elements. To assess the reliability of the program's cost estimate, we compared agency documentation of the program office estimate and the independent cost estimate, and interviewed program officials and cost estimators to understand key aspects of and differences between the estimates. We determined that the estimates were sufficient for our purposes of providing summary data. We interviewed program officials from NOAA, DOD, and the National Aeronautics and Space Administration (NASA), to obtain information on transition schedules, progress, program requirements, and challenges.

To assess NOAA's progress in developing the NPP and JPSS satellite systems, we compared NOAA's plans for key milestones to its actual accomplishments. We reviewed monthly progress reports, draft program schedules, and the NPP operational readiness review package. We observed NOAA's monthly program management briefings to determine the status of key components. We interviewed both agency and contractor officials, including officials at Ball Aerospace, Inc. and Raytheon Space and Airborne Systems, Inc. We also interviewed key NOAA satellite data users, including officials involved in weather forecasting and numerical weather prediction, to identify their experiences in working with NPP data as well as their plans for working with JPSS data.

To evaluate NOAA's efforts to mitigate key project risks, we compared the
agency's risk management process to best practices in risk management
as identified by the Software Engineering Institute. We reviewed NOAA's
program risk lists on a monthly basis to obtain insights into management
issues and actions. We interviewed agency and contractor officials to
evaluate actions to address each transition risk. In addition, we
interviewed NOAA satellite data users to determine the impact of any
changes in requirements.

We performed our work at NASA, NOAA, and DOD offices in the
Washington, D.C., area and at contractor facilities in Los Angeles,
California; Aurora, Colorado; and Boulder, Colorado. We conducted this
performance audit from May 2011 to June 2012 in accordance with
generally accepted government auditing standards. Those standards
require that we plan and perform the audit to obtain sufficient, appropriate
evidence to provide a reasonable basis for our findings and conclusions
based on our audit objectives. We believe that the evidence obtained
provides a reasonable basis for our findings and conclusions based on
our audit objectives.

Appendix II: Comments by the Department of Commerce

UNITED STATES DEPARTMENT OF COMMERCE
The Secretary of Commerce
Washington, D.C. 20230

May 16, 2012

Mr. David A. Powner
Director, Information Technology Management Issues
U.S. Government Accountability Office
441 G Street NW
Washington, DC 20548

Dear Mr. Powner:

Thank you for the opportunity to review and comment on the Government Accountability Office's draft report entitled *Polar-Orbiting Environmental Satellites: Changing Requirements, Technical Issues, and Looming Data Gaps Require Focused Attention* (GAO-12-604). On behalf of the Department of Commerce, I have enclosed the National Oceanic and Atmospheric Administration's programmatic comments to the draft report.

If you have any questions, please contact Jim Stowers, Acting Assistant Secretary for Legislative and Intergovernmental Affairs at (202) 482-3663.

Sincerely,

John E. Bryson

Enclosure

Department of Commerce
National Oceanic and Atmospheric Administration
Comments to the Draft GAO Report Entitled
"Polar-orbiting Environmental Satellites: Changing Requirements, Technical
Issues, and Looming Data Gaps Require Focused Attention"
(GAO-12-604, June 2012)

<u>General Comments</u>

The Department of Commerce's National Oceanic and Atmospheric Administration (NOAA) appreciates the opportunity to review the Government Accountability Office's (GAO) draft report on the Polar-orbiting Environmental Satellites. The draft report on Polar-Orbiting Environmental Satellites requirements, technical issues, and potential gaps does a fair job of assessing the state of the program. Given the history and changes in the polar-orbiting environmental satellite programs over the last decade, a number of recommended changes and updates have been provided that will ensure that the information presented is complete, up-to-date, and reflects the most current plans.

<u>NOAA Response to GAO Recommendation</u>

Recommendation: "Given the importance of polar-orbiting satellite data to weather forecasts, we recommend that the Secretary of Commerce direct the Administrator of NOAA to establish mitigation plans for risks associated with pending satellite data gaps in the afternoon orbit as well as potential gaps in the morning and midmorning orbits."

NOAA Response: NOAA agrees with this recommendation. The National Environmental Satellite, Data and Information Service (NESDIS) has performed these analyses, but have not compiled them in a single report for formal review and approval. NESDIS plans to finalize this report by August 31, 2012 for NOAA approval.

1

Appendix III: GAO Contact and Staff Acknowledgments

GAO Contact	David A. Powner (202) 512-9286 or pownerd@gao.gov
Staff Acknowledgments	In addition to the contact named above, Colleen Phillips (Assistant Director), Kathleen Lovett Epperson, Kate Feild, Nancy Glover, Franklin Jackson, and Fatima Jahan made key contributions to this report.

Related GAO Products

NASA: Assessments of Selected Large-Scale Projects. GAO-12-207SP. (Washington, D.C.: Mar. 1, 2012).

Polar Satellites: Agencies Need to Address Potential Gaps in Weather and Climate Data Coverage. GAO-11-945T. (Washington, D.C.: Sept. 23, 2011).

Polar-Orbiting Environmental Satellites: Agencies Must Act Quickly to Address Risks That Jeopardize the Continuity of Weather and Climate Data. GAO-10-558. (Washington, D.C.: May 27, 2010).

Polar-Orbiting Environmental Satellites: With Costs Increasing and Data Continuity at Risk, Improvements Needed in Tri-agency Decision Making. GAO-09-772T. (Washington, D.C.: June 17, 2009).

Polar-orbiting Environmental Satellites: With Costs Increasing and Data Continuity at Risk, Improvements Needed in Tri-agency Decision Making. GAO-09-564 (Washington, D.C.: June 17, 2009).

Environmental Satellites: Polar-orbiting Satellite Acquisition Faces Delays; Decisions Needed on Whether and How to Ensure Climate Data Continuity. GAO-08-899T. (Washington, D.C.: June 19, 2008).

Environmental Satellites: Polar-orbiting Satellite Acquisition Faces Delays; Decisions Needed on Whether and How to Ensure Climate Data Continuity. GAO-08-518. (Washington, D.C.: May 16, 2008).

Environmental Satellite Acquisitions: Progress and Challenges. GAO-07-1099T. (Washington, D.C.: July 11, 2007).

Polar-orbiting Operational Environmental Satellites: Restructuring Is Under Way, but Challenges and Risks Remain. GAO-07-910T. (Washington, D.C.: June 7, 2007).

Polar-orbiting Operational Environmental Satellites: Restructuring Is Under Way, but Technical Challenges and Risks Remain. GAO-07-498. (Washington, D.C.: Apr. 27, 2007).

Polar-orbiting Operational Environmental Satellites: Cost Increases Trigger Review and Place Program's Direction on Hold. GAO-06-573T. (Washington, D.C.: Mar. 30, 2006).

Polar-orbiting Operational Environmental Satellites: Technical Problems, Cost Increases, and Schedule Delays Trigger Need for Difficult Trade-off Decisions. GAO-06-249T. (Washington, D.C.: Nov. 16, 2005).

Polar-orbiting Environmental Satellites: Information on Program Cost and Schedule Changes. GAO-04-1054. (Washington, D.C.: Sept. 30, 2004).

Polar-orbiting Environmental Satellites: Project Risks Could Affect Weather Data Needed by Civilian and Military Users. GAO-03-987T. (Washington, D.C.: July 15, 2003).

Polar-orbiting Environmental Satellites: Status, Plans, and Future Data Management Challenges. GAO-02-684T. (Washington, D.C.: July 24, 2002).

GAO's Mission	The Government Accountability Office, the audit, evaluation, and investigative arm of Congress, exists to support Congress in meeting its constitutional responsibilities and to help improve the performance and accountability of the federal government for the American people. GAO examines the use of public funds; evaluates federal programs and policies; and provides analyses, recommendations, and other assistance to help Congress make informed oversight, policy, and funding decisions. GAO's commitment to good government is reflected in its core values of accountability, integrity, and reliability.
Obtaining Copies of GAO Reports and Testimony	The fastest and easiest way to obtain copies of GAO documents at no cost is through GAO's website (www.gao.gov). Each weekday afternoon, GAO posts on its website newly released reports, testimony, and correspondence. To have GAO e-mail you a list of newly posted products, go to www.gao.gov and select "E-mail Updates."
Order by Phone	The price of each GAO publication reflects GAO's actual cost of production and distribution and depends on the number of pages in the publication and whether the publication is printed in color or black and white. Pricing and ordering information is posted on GAO's website, http://www.gao.gov/ordering.htm. Place orders by calling (202) 512-6000, toll free (866) 801-7077, or TDD (202) 512-2537. Orders may be paid for using American Express, Discover Card, MasterCard, Visa, check, or money order. Call for additional information.
Connect with GAO	Connect with GAO on Facebook, Flickr, Twitter, and YouTube. Subscribe to our RSS Feeds or E-mail Updates. Listen to our Podcasts. Visit GAO on the web at www.gao.gov.
To Report Fraud, Waste, and Abuse in Federal Programs	Contact: Website: www.gao.gov/fraudnet/fraudnet.htm E-mail: fraudnet@gao.gov Automated answering system: (800) 424-5454 or (202) 512-7470
Congressional Relations	Katherine Siggerud, Managing Director, siggerudk@gao.gov, (202) 512-4400, U.S. Government Accountability Office, 441 G Street NW, Room 7125, Washington, DC 20548
Public Affairs	Chuck Young, Managing Director, youngc1@gao.gov, (202) 512-4800 U.S. Government Accountability Office, 441 G Street NW, Room 7149 Washington, DC 20548